BLACKTIP REEF SHARKS

by Julie K. Lundgren

A Crabtree Seedlings Book

TABLE OF CONTENTS

Crabtree Publishing
crabtreebooks.com

School-to-Home Support for Caregivers and Teachers

This book helps children grow by letting them practice reading. Here are a few guiding questions to help the reader with building his or her comprehension skills. Possible answers appear here in red.

Before Reading:

- What do I think this book is about?
    - *I think this book is about blacktip reef sharks.*
    - *I think this book is about where they live.*

- What do I want to learn about this topic?
    - *I want to learn about the habits of blacktip reef sharks.*
    - *I want to learn if it's safe to swim with them.*

During Reading:

- I wonder why...
    - *I wonder why they have black on the tip of their fins.*
    - *I wonder why they hear and see well.*

- What have I learned so far?
    - *I have learned that blacktip reef sharks live near reefs in oceans.*
    - *I have learned that sharks lose and grow new teeth all their lives.*

After Reading:

- What details did I learn about this topic?
    - *I have learned that divers can safely swim with these sharks.*
    - *I have learned that a reef is a shallow, underwater ridge where many sea animals live.*

- Read the book again and look for the vocabulary words.
    - *I see the word **oceans** on page 4 and the word **divers** on page 18. The other glossary words are on pages 22 and 23.*

BLACKTIP REEF SHARKS

Meet the blacktip **reef** shark.

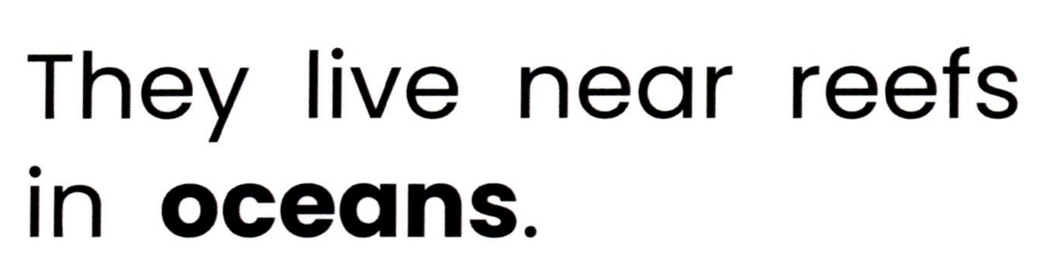

They live near reefs in **oceans**.

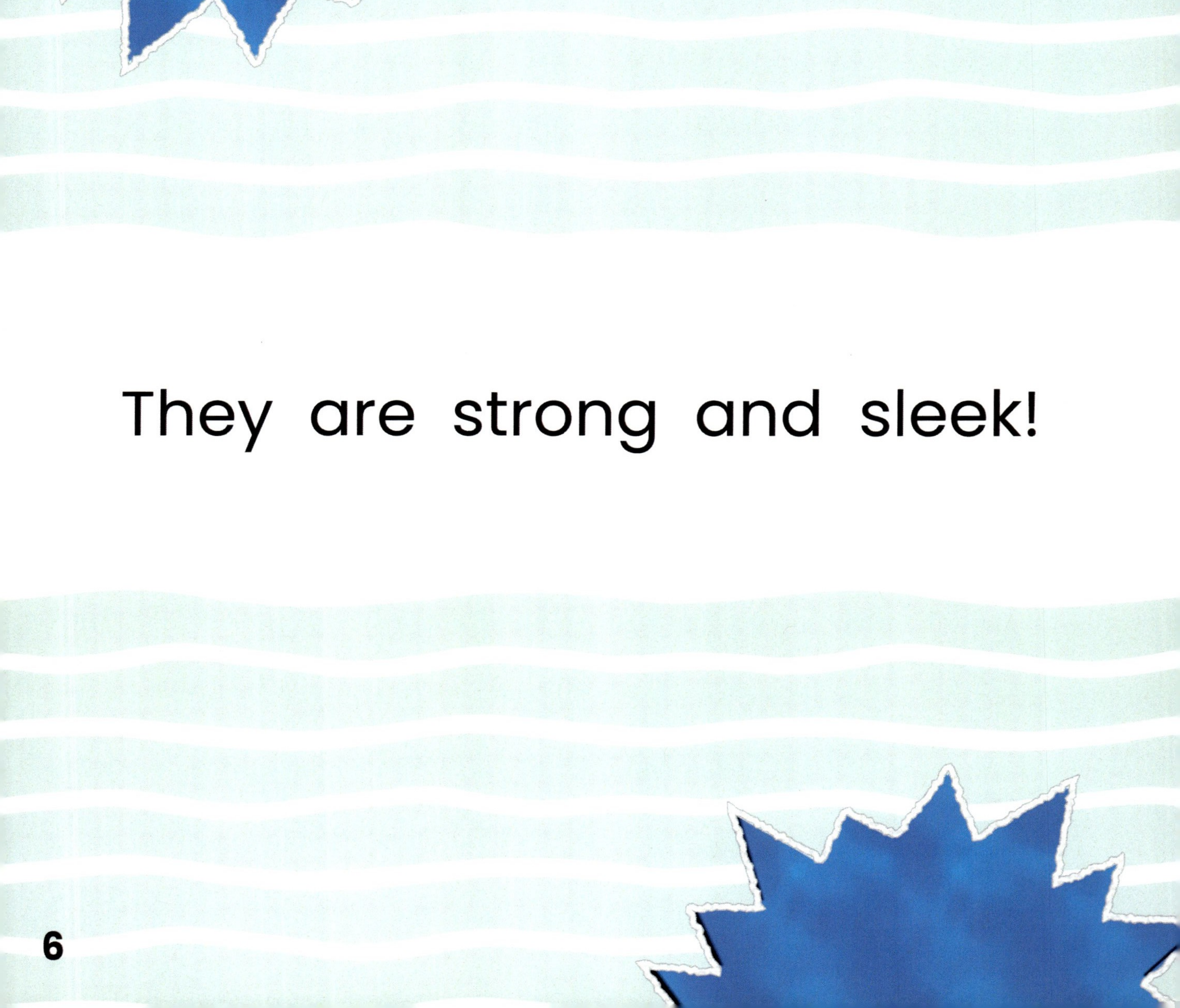

They are strong and sleek!

FROM THE FILES

The average blacktip reef shark grows to about 5.5 feet (1.7 m) long.

See the black tip on the top **fin**?

fin

Fins help them **balance** and steer.

FROM THE FILES

Surfboard fins steer, too!

Sharp **teeth** help them catch fish to eat.

FROM THE FILES

Sharks lose and grow new teeth all their lives.

They see and hear well.

They may swim together
or alone.

Divers can safely swim with these sharks.

Would you swim with
blacktip reef sharks?

GLOSSARY

balance (BAL-uhnss): To balance is to keep steady and not tip over.

divers (DIE-verz): Divers are people that wear gear to help them breathe underwater.

fin (FIN): A fin is a part on a fish's body that is used to hel the fish steer and balance.

oceans (OH-shuhnz): Oceans are large bodies of saltwater where many animals live.

reef (REEF): A reef is a shallow, underwater ridge where many sea animals live.

teeth (TEETH): Teeth are white, bony parts inside a mouth that are used for biting and chewing.

Index

About the Author

Julie K. Lundgren

Julie K. Lundgren grew up near Lake Superior where she reveled in mucking about in the woods, picking berries, and expanding her rock collection. Her interests led her to a degree in biology. She lives in Minnesota with her family.

Websites

https://aqua.org/explore/exhibits/blacktip-reef
www.montereybayaquarium.org/animals/animals-a-to-z/blacktip-reef-shark

---

Crabtree Publishing

crabtreebooks.com 800-387-7650

Written by: Julie K. Lundgren
Designed by: Jennifer Dydyk
Edited by: Kelli Hicks
Proofreader: Melissa Boyce

Photographs:
Shark illustration on cover logo © BATKA/Shutterstock; Cover photo © Ian Scott/Shutterstock, page 3 © Gino Santa Maria/Shutterstock; page 5 © Kristina Vackova/ Shutterstock; page 7 © antos777/Shutterstock, diver © Ian Scott/ Shutterstock, shark illustration © Dashikka/Shutterstock; page 9 © Ian Scott/ Shutterstock; page 11 © Sergey Utkin/Shutterstock, surfer © Wonderful Nature, shark illustration © Dashikka/Shutterstock; page 12 © Karel Bartik/Shutterstock; Page 13 © Mark_Kostich/Shutterstock, shark illustration © Dashikka/ Shutterstock; page 15 © Yann hubert/Shutterstock; Page 17 © Tomas Kotouc/ Shutterstock; page 19 © Ian Scott/Shutterstock; page 21 © Dray van Beeck/ Shutterstock; page 23 reef photo © Richard Whitcombe/Shutterstock

Hardcover	978-1-4271-5824-6
Paperback	978-1-4271-5825-3
Ebook (pdf)	978-1-4271-5826-0
Epub	978-1-4271-5827-7
Read-along	978-1-4271-5828-4
Audio book	978-1-4271-5829-1

Printed in Canada/102024/CP20241003

Published in Canada
Crabtree Publishing
616 Welland Avenue
St. Catharines, Ontario
L2M 5V6

Published in the United States
Crabtree Publishing
347 Fifth Avenue
Suite 1402-145
New York, NY 10016

Library and Archives Canada Cataloguing in Publication

Available at the Library and Archives Canada

Library of Congress Cataloging-in-Publication Data

Names: Lundgren, Julie K., author.
Title: Blacktip reef sharks / Julie K. Lundgren.
Description: New York : Crabtree Publishing, [2022] | Series: Shark files - a Crabtree seedlings book | Includes index.
Identifiers: LCCN 2021018433 (print) | LCCN 2021018434 (ebook) | ISBN 9781427158246 (hardcover) | ISBN 9781427158253 (paperback) | ISBN 9781427158260 (ebook) | ISBN 9781427158277 (epub) | ISBN 9781427158284
Subjects: LCSH: Blacktip shark--Juvenile literature.
Classification: LCC QL638.95.C3 L863 2022 (print) | LCC QL638.95.C3 (ebook) | DDC 597.3/4--dc23
LC record available at https://lccn.loc.gov/2021018433
LC ebook record available at https://lccn.loc.gov/2021018434